Do Birds Sneeze?

For information contact: Jennifer Kuhns, Author/Publisher
http://www.jenniferkuhns.net

ISBN 979-8-9877485-0-3

Cover design: Jennifer Kuhns
Formatting: Jennifer Kuhns
Cover Art and Illustrations: Steven Kistler

PRINTED IN THE UNITED STATES OF AMERICA

Do Birds Sneeze?

Can worms see?

Earthworms do not have eyes, so they cannot see. What they do have are something called light-sensitive cells that are located throughout the outer layer of their skin. Earthworms cannot use these cells to see, but they help worms detect light and changes in light brightness or intensity. These light-sensitive cells are also sensitive to touch and chemicals like things called pesticides or insecticides.

Do lizards jump?

Yes, lizards can jump, but their body is not actually made for jumping. Jumping is not their best or strongest ability. They prefer to run instead of jumping. Lizards can jump from a high place to a lower one to get to where they want to go for hunting and moving around but when threatened or scared, lizards run. So, they don't jump up, but they can jump down.

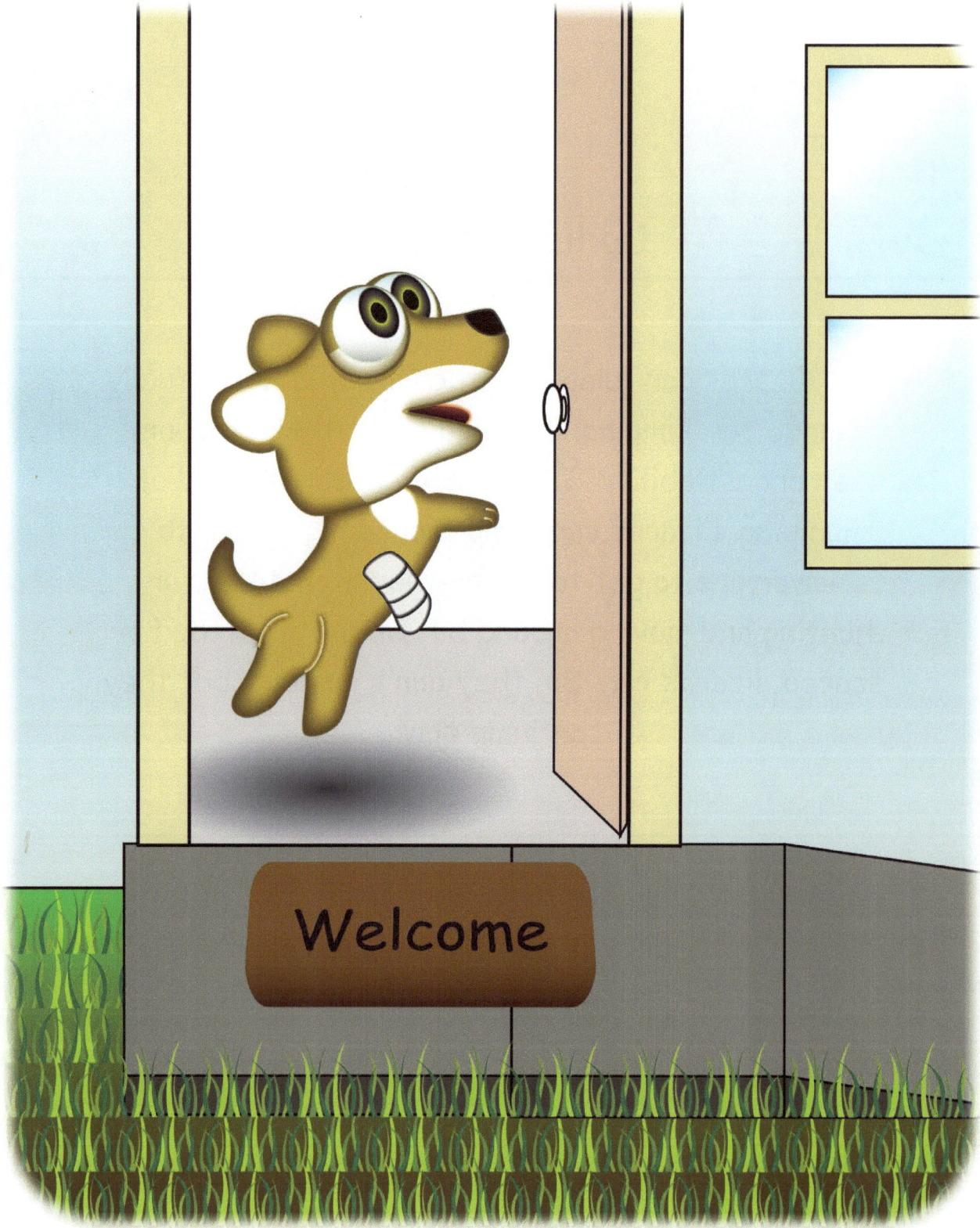

Welcome

Can dogs skip?

Most of the time, dogs skip when they walk because their legs are stiff, usually from the cold. Most dogs will get stiff after spending time outside in the cold or laying on cold ground, so their first few steps will look like they are skipping. They will then settle back into a normal way of walking after a few seconds of getting warm. Another reason could be something serious like an injury or pain from arthritis. If your dog is otherwise healthy, she probably won't have any aches and pains that are serious enough to cause her to skip.

Can fish drown?

This may be a surprise, but fish can drown. They can easily die from a lack of oxygen. Believe it or not, fish require oxygen to survive just like humans and air-breathing animals. In fact, all living things, fish and plants included, use oxygen to break down food that releases energy. That energy is needed and used to keep all things alive. When fish don't have air or oxygen, it's impossible for them to stay alive.

Do birds sneeze?

Yes, birds have a habit of sneezing once or twice a day. It is their daily habit, and it helps them clear their air passage of dust and other bothersome things. Sometimes they even have clear snot! If you witness this, nothing is wrong and there is nothing to worry about.

However, just like you, a bird can have trouble breathing if they are sick. If a bird is sick in any way, then you will notice the bird sneezing more often, and they would also have thick slightly colored snot. The same goes for coughing.

Can snakes hear?

The truth is that snakes do hear but not in the same way humans do. With humans, the sound wave travels through the air and hit the eardrum and causes vibrations in the tiny hair cells and the movement of small bones located in the inner ear. These vibrations change into nerve impulses that travel to the brain.

Although snakes have small holes located on the sides of their heads, which are the ear openings, they do not have eardrums like humans. Instead, sound waves in the air have a shorter trip. The snake's skulls vibrate once the sound waves hit their heads. That message travels directly from the snake's skull to its inner ears, where the brain senses the vibrations.

Do plants feel?

Yes, plants really can *feel* when they are being touched. A good example would be the plant called Venus Flytrap. When a fly lands on one of its leaves, it snaps shut, trapping the insect, which is then eaten by the plant. What's even more curious is that a Venus Flytrap can tell the difference between an insect and a drop of rain. It won't snap closed when a drop of water hits it.

Do bears really hug?

Yes, a bear hugs, but a "bear-hug" is when a bear hugs his prey tightly with both hands and squeezes him to death. (Sometimes mom's hug almost as hard because they love you.)

Do spiders fart?

Surprisingly, the answer is yes. Spiders actually have a digestive system and anus just like we do, which means they do fart!

Do crabs have teeth?

Yes, as strange as it may seem, crabs do have teeth, but not in their mouths. WHAT! Crabs have teeth in their stomachs that are used to chomp up their food before they digest it.

So how do they get food from their mouths into their stomach without chewing it? Well, depending on the species of crab, they may have soft or hard mouthparts that help move food into their mouth. For example, the Dungeness crab has several types of mouthparts, sort of like our jaws, that are used for holding food, for breaking it up, and others for moving food further into their mouth. Some species such as the Ghost crab even use their stomach teeth to growl and warn off predators.

Can humming birds walk?

Hummingbirds may have feet but they are not used for walking. They are only able to use their feet for preening, scratching and perching. At best, they can slowly shuffle. If you see a hummingbird on the ground, it is likely because they are in trouble.

Hummingbirds have four toes; three toes in the front and one toe (called the hallux) on the back of the foot. The hallux works in the same way a human's thumb does and allows the hummingbird to hang on to a branch or wire. Because hummingbirds can fly backwards and hover in the air, they have evolved to have smaller, lighter feet than other birds. That makes hummingbirds more aerodynamic and better at flying.

Is the sky really blue?

Well, let's figure this out. Light travels from the sun in waves and is made up of many colors. Blue is a short wavelength; red is long and all the others are in between. By the time the light gets to us here on earth, it had to pass through and bend around a lot of things, so it gets scattered. It's the waves that are shorter that get through to us. So that's why the short blue ones get through, but the longer ones don't. So, is the sky really blue?

Does money grow on trees?

This is something called a proverb. A proverb is a short saying that has a little bit of truth to it. Real money does not grow on trees. So, if someone runs out of money, they cannot just go out to their backyard and pull dollar bills off the branches of a tree. But. . . a person can pick fruit off of a tree to sell and make money. No, money is not so easy to get. A person has to work hard for it.

Why does it rain?

When the air holds lots of water droplets, clouds form. If a lot of water droplets gather in the clouds, the clouds become heavy. Gravity causes the water droplets to fall as rain. This is how rain is formed.

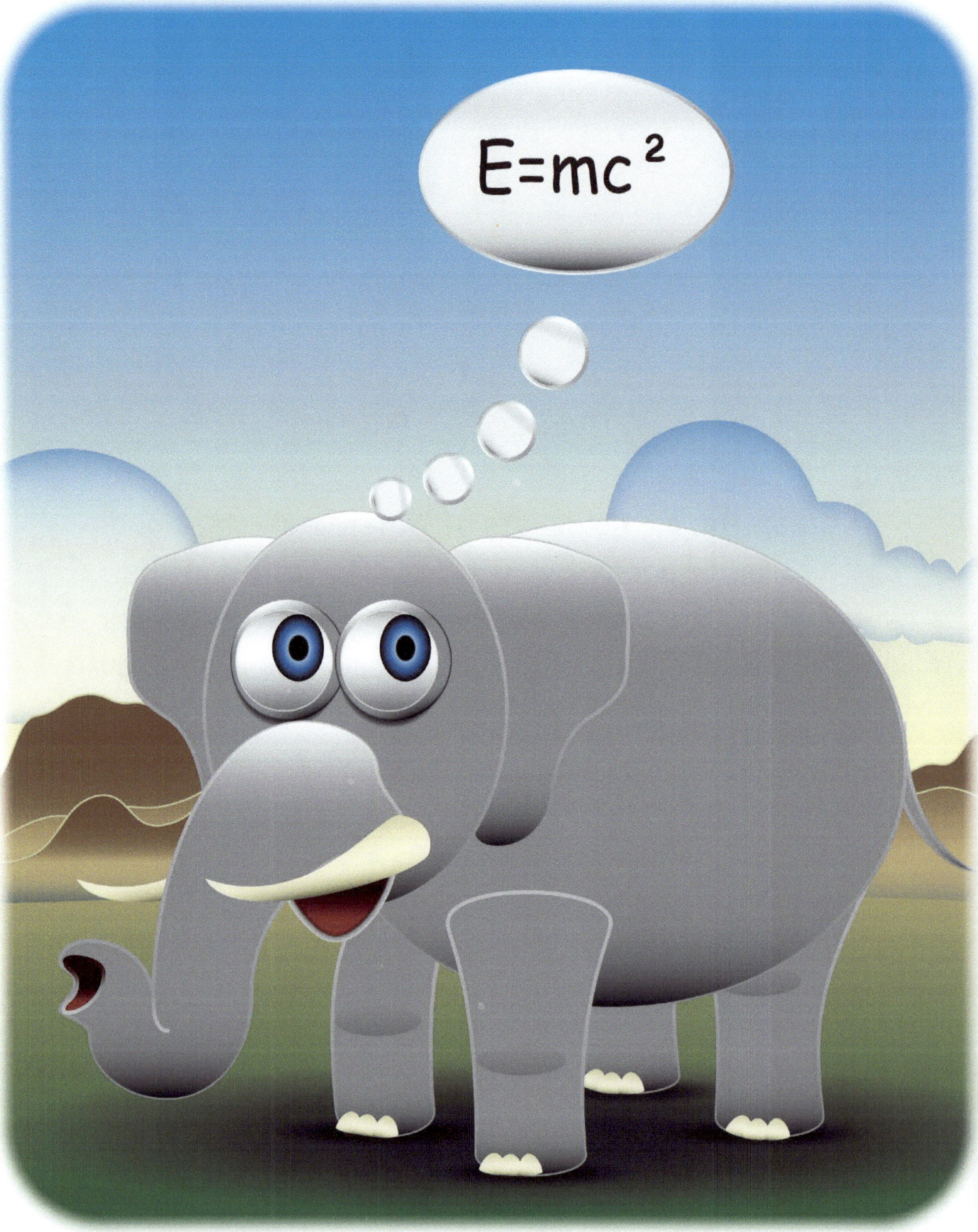
E=mc ²

Does an elephant really never forget?

In reality, "an elephant never forgets" is not true all the time because all elephants forget things from time to time. However, scientists have proven that elephants do have incredible memories. They seem to remember other elephants when they see them. They also seem to have something called recall power that helps keep them safe and away from danger. That's why elephants live longer than other animals. As they travel, the elephants have the ability to remember where things like watering holes are and the best food.

Why do bees buzz?

When we hear bees buzzing it is because we hear their tiny wings beating really, really fast. Some bees beat their wings as many as 230 times a SECOND. That makes quite a buzzing sound. This is also why we hear buzzing sounds from flies, wasps, and all kinds of other fast-flying insects.

Do dogs laugh?

Yes! Dogs do laugh; however, it is not the same way humans do.

In humans, laughter is made up of things called rhythmic, vocalized, expiratory, and involuntary actions. Human laughing sounds can be any variation of "ha-ha" or "ho-ho." Dogs, on the other hand, make sort of a similar sound when they pant hard or huff—like a "hhuh-hhah" noise. Dogs usually make this sound while playing or trying to invite humans and other dogs to play; it is known as a "play-pant." So, play-pant is a form of breathing and not really a vocal sound like humans make.

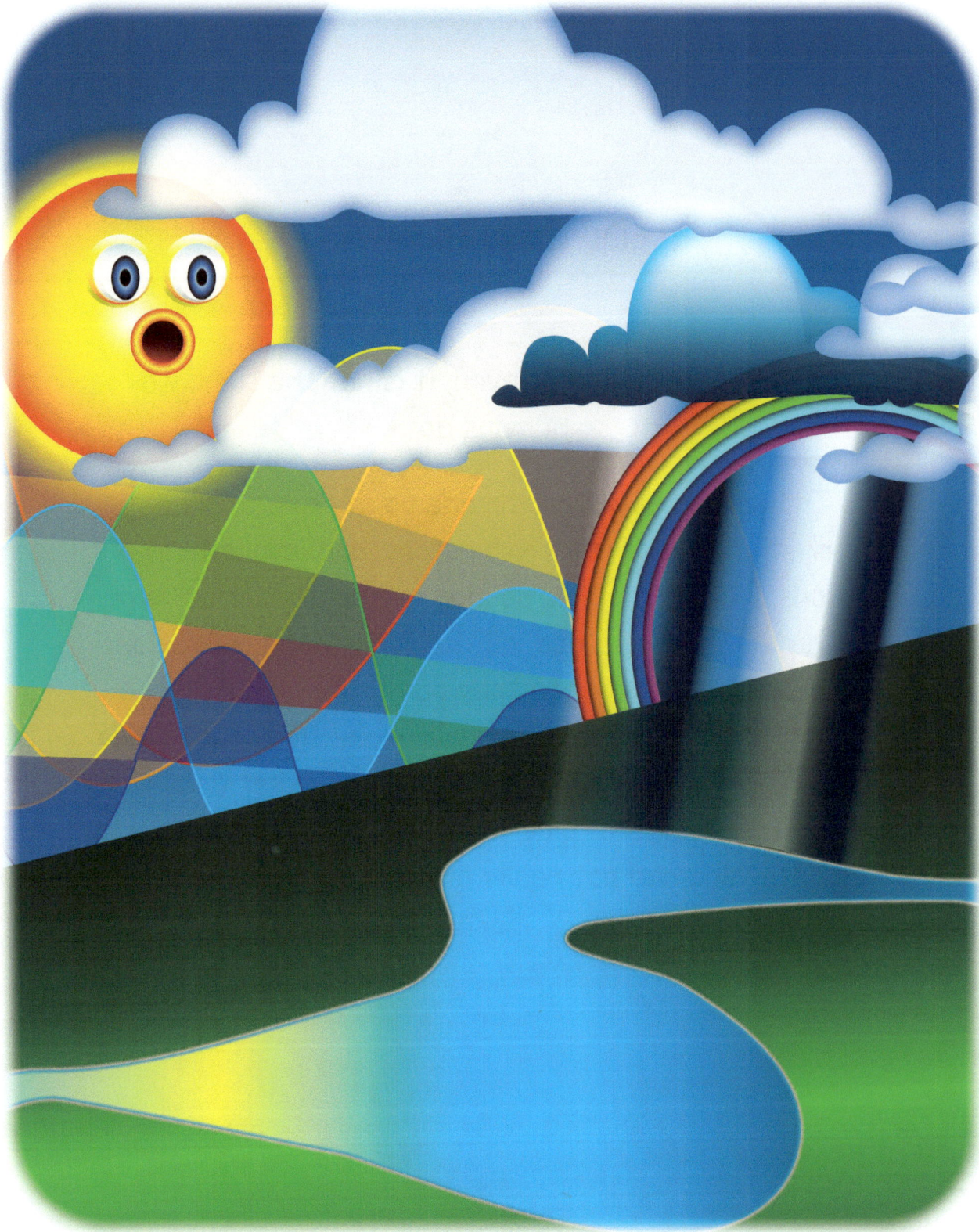

What is at the end of a rainbow?

Light from the sun looks white, but it is actually made of lots of different colors. Normally, they are all mixed together (remember why the sky is blue?). A rainbow is formed when light from the sun meets raindrops in the air and the raindrops separate out all these different colors.

Because rainbows are made in the sky, they don't touch the ground. So, if you're on the ground, however far you walk, the end of the rainbow will always look far away.

But what people don't realize is that rainbows are actually complete circles, and obviously a circle has no end. You just never see the whole circle because the earth's horizon gets in the way.

So, what is at the end of the rainbow: Nothing.

Why am I left-handed?

Scientists aren't exactly sure why some people are left-handed, but they know that genes or genetics (who your mom and dad are and if they are left-handed) are responsible for about 25% of the reason you are left-handed. Scientists do know that Left-handedness does tend to run in families. There are lots of theories on what else might determine which hand you write with, but many experts believe that it's really kind of random. Some scientists believe the brain decides for you and other scientists think the answer is in the spinal cord. They all agree it happens before you are born.

Why don't veggies taste better?

Vegetables taste bad because of something called a defense mechanism. Yes, it is true that vegetables have their own defense mechanism. Have you ever wondered why your mom cries when she is cutting onions? It is because the onions are fighting back. Just like onions, when you start cutting other vegetables, something called mustard oil, a natural chemical that is part of the vegetables, is released. The release of the mustard oil is a self-defense mechanism, which makes the vegetables taste a little bitter. These natural chemicals most likely help the vegetable plant to defend itself against pests and diseases because bugs don't like the bitter taste either. Vegetables actually keep fighting back until you completely chew them. So, here's the answer to the equation- Does mustard oil make the vegetables taste bad? ...Yes

Where does the sun go when it gets dark?

At night, the sun does not go anywhere. Day or night, the sun stays exactly at the center of our solar system where it has always been, and will be for the next bazillion years. What makes the sun look like it disappears when it gets dark is because of the Earth's rotation or turning motion . . .the sun really and truly never goes anywhere. It actually stays in the center of our solar system. Even though we call it sunrise when it gets light in the morning and sunset when it gets dark at night, it's really all about the earth's rotation that causes us to get up each morning and go to bed each night.

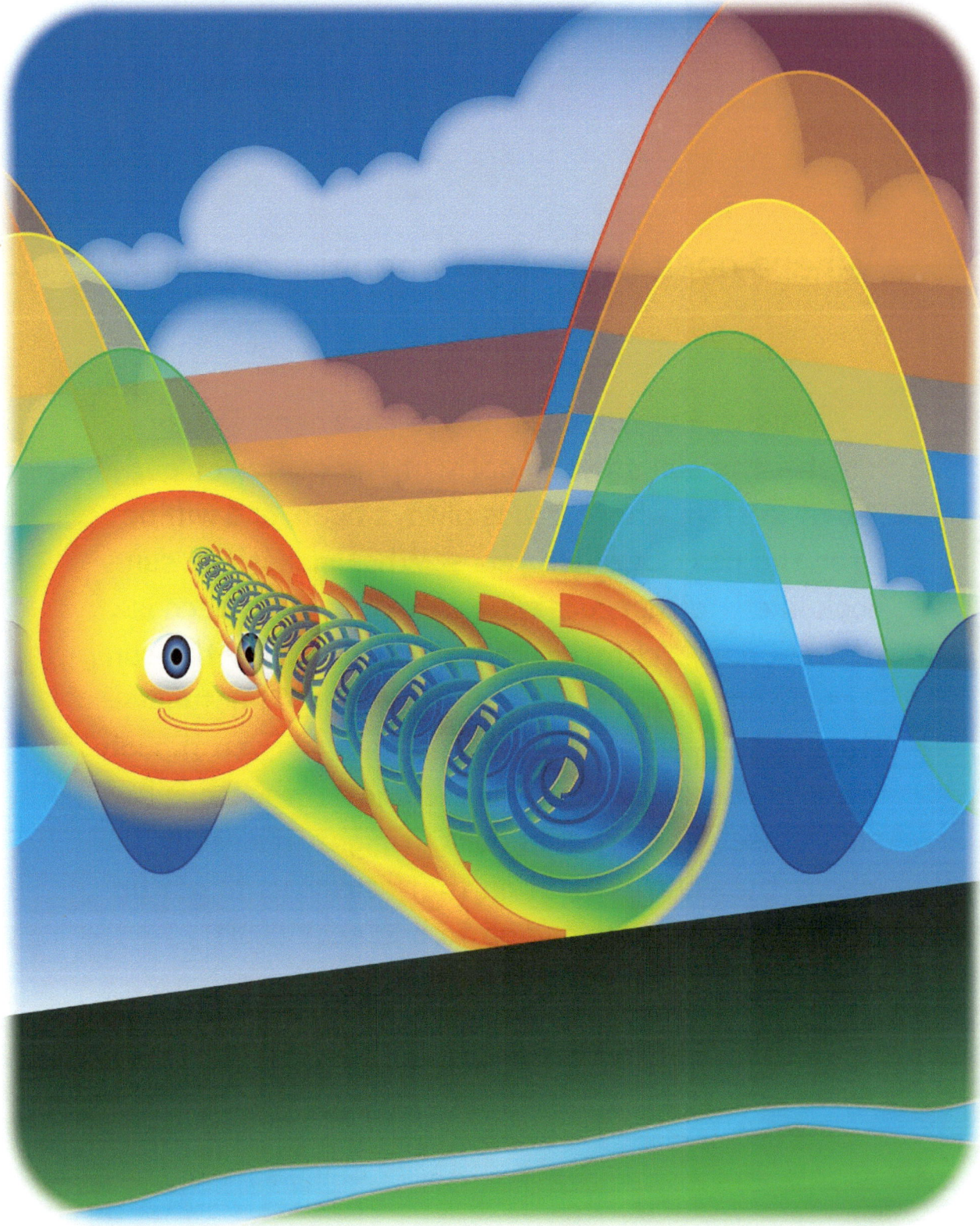

What is light made of?

Light is made from something called **electromagnetic energy**. This kind of energy is formed from burning gases. The sun burns gasses to emit (or give off) light. When the sun's gasses burn it causes a chemical reaction that makes this thing called **electromagnetic energy**, which is a type of energy that takes the form of magnetic and electrical waves. Waves are the invisible vibrations that affect objects and even space to create light, color and more. The electromagnetic energy created from the burning gasses is what makes light.

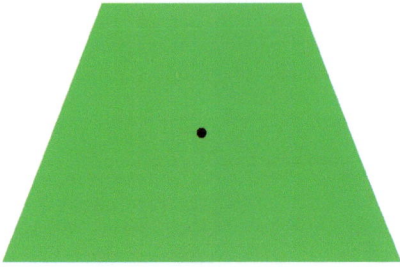

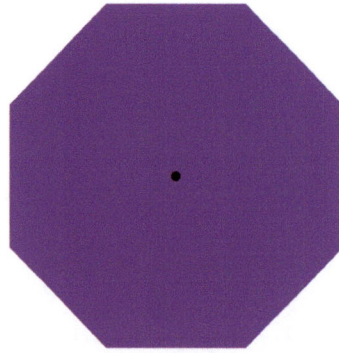

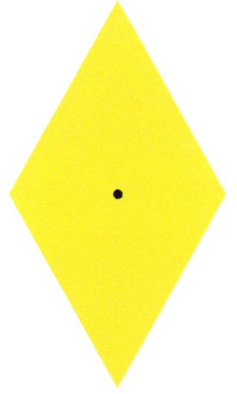

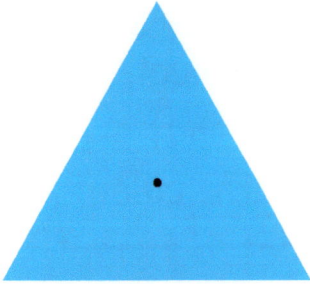

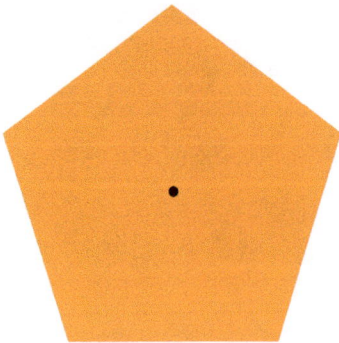

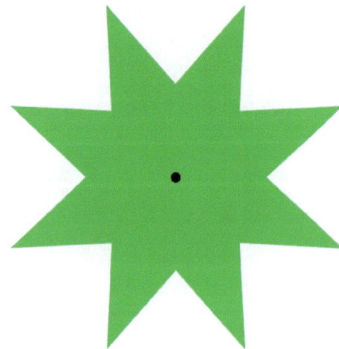

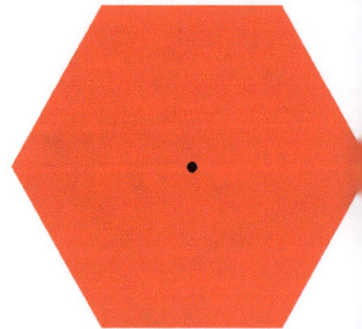

Where does a circle end?

Nowhere... A Circle is a two-dimensional shape that is made from one continuous line . . .and the center point of a circle, the middle, is the same distance away from that line all the way around the circle.

More children's books by Jennifer Kuhns

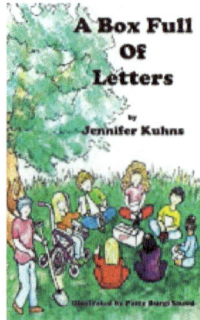

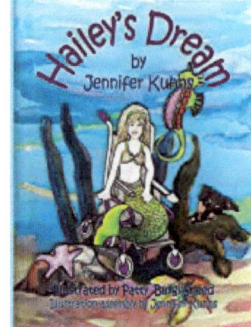

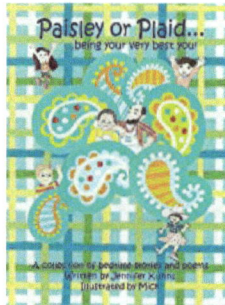

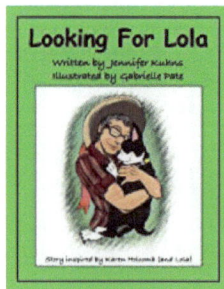

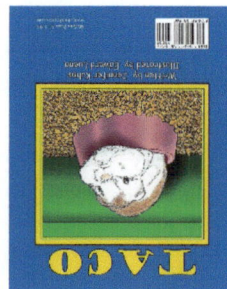

WERE YOU BORN IN THAT CHAIR?
by
Jennifer Kuhns

Illustrated by Karen Borrelli

A Box Full
Of
Letters
by
Jennifer Kuhns

Illustrated by Patty Borgo Sound

Hailey's Dream
by
Jennifer Kuhns

Illustrated by Patty Borgo Sound
Illustrations assembly by Jennifer Kuhns

Paisley or Plaid...
being your very best you!

A collection of peppish stories and poems
Written by Jennifer Kuhns
Illustrated by MCK

Looking For Lola
Written by Jennifer Kuhns
Illustrated by Gabrielle Pate

Story inspired by Karen Holcomb (and Lola)

TACO
Written by Jennifer Kuhns
Illustrated by Patty Borgo Sound

A two-sided flipbook

Miles to the Moon

Written by
Jennifer Kuhns
Illustrated by
Edward Luena

Lilly Gets Lucky

Written and illustrated by
Josephine Nail
and
Jennifer Kuhns

Supporting illustrator
Mike Kuhns

Most People Don't Talk To Rocks

Jennifer Kuhns
Illustrated by Mikai Kuhns

My Grandma Taught Me How To Throw Stuff

by
Jennifer Kuhns
Illustrated by
Linda Oyler

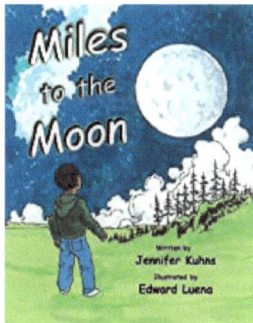

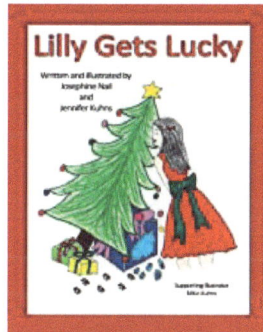

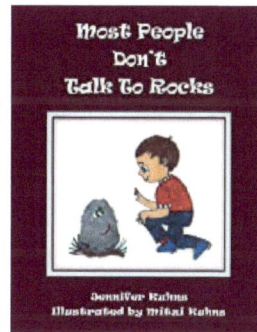

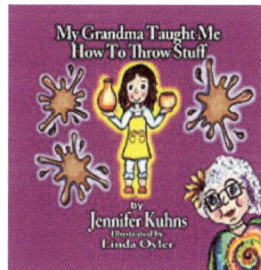

For the adult reader

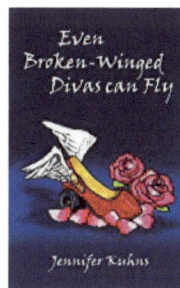

Jennifer Kuhns
Little Diva On Wheels

Growing Up Differently-Abled

Even Broken-Winged Divas can Fly

Jennifer Kuhns